The Structure of Everything

The Structure of Everything

A New String Theory
or "The Music of the Spheres."

James Byron Walker

Copyright © 2009 by James Byron Walker.

ISBN: Softcover   978-1-4415-0698-6

All rights reserved. No part of this book may be reproduced or transmitted in any form or by any means, electronic or mechanical, including photocopying, recording, or by any information storage and retrieval system, without permission in writing from the copyright owner.

This book was printed in the United States of America.

To order additional copies of this book, contact:
Xlibris Corporation
1-888-795-4274
www.Xlibris.com
Orders@Xlibris.com

Contents

Chapter 1: Introduction of Quantum construct concept 7
Angular Momentum in point particles explanation
Inter-related co-variants, not invariants, explanation

Chapter 2: Others ideas and their unseen implication 13
$E = MC$ squared + implications
Curvature of space-time + implications
# of Quanta / electron + how I arrived at it
Refutation of present theories of size and age of Universe
Mass of Big Bang, Min + MAX., 2 methods of arriving at Refutation of Natl. Geographic Oct. 1999 article on Universes expansion
Possible time dilation equation for acceleration proposed

Chapter 3: Introduction of Fractals, Mandelbrot sets & Julia sets......19
Possible 4 dimension Julia equations proposed

Chapter 4: Definition of Quanta, The Paradox of light
& consequences 22
Explanation of phenomena of strings

Chapter 5: Time & Entropy, Explanation 27

Chapter 6: 2 possible explanations for the neutrino & analogy........28

Chapter 7: Possible shapes of the Universe + consequences............30

Chapter 8: Problems determining if we are
in a closed event horizon............33
Universe + consequences
Paradoxes of an event horizon+ implications & consequences

Chapter 9:   Big Bang misnomer, explanation + consequences..........36
             Other possible naked singularities

Chapter 10: The Evolution of galaxies 39

Chapter 11: Genetics...42
             Genetic ring of Species, explanation
             9 methods of genetic change, explanation
             Introduction of G.L.A.S.S.E.S. concept
             Amino Acid evolution- explanation of processes

Chapter 12: Form follows function $\{f(x) = x-u\}$ 50
             Introduction of Evolving Julia sets as the final solution

Chapter 1

Introduction of Quantum construct concept

Modern Physics is much too complicated; Alternative Universes, branes, dark matter, strings, quarks, gluons, Up down and strange particles, Saddle Shaped Universes, the Universe expanding at an increasing pace and other such conundrums. There is no explanation for the structure of anti-matter except that it annihilates matter on contact; there is no explanation as to why, or its structure, no real understanding of string theory except that it only works in 10 dimensions to seventeen dimensions, with no explanation as to why. There is no real understanding of mass, or gravity, magnetic force, the weak atomic force, the strong atomic force or other forces. They, the physicists, try to find connections and relationships between these forces and can't find them. The last significant work was done in this field by Einstein in 1915. No Quantum structure of the atom has been proposed since Niels Bohr and I will attempt to show how it is outdated and doesn't fit the parameters of reality. How can you have an understanding of Quantum physics and mechanics when you don't have a Quantum structure of the atom? The obvious answer is you can't.

I first came up with this theory more than 45 years ago during the fall of 1961 and the first part of 1962 in my senior year of high school. Then, the theory, actually a quantum construct of the atom, only included the structure of the atom, including matter, anti-matter, neutrinos, and all of the particles and sub-atomic particles known and theorized up until that time. I also predicted the charge, spin, and mass of four particles before they were found. I don't remember the exact details, but I do remember predicting the mass, spin, charge and weight of four particles in my senior year of high school. I presented it to my physics teacher, and he couldn't even grasp the basics of what I was trying to do, much less help with some problems I was having about where energy added to a system went.

At that point I dropped my investigations and ruminations on the subject as going nowhere. It wasn't until years later that I figured where the added energy went. This theory, or construct if you will, explains the charges, spins, and mass of all of these including several possible explanations of neutrinos. It also included the structure of anti-matter and an explanation as to why and how anti-matter annihilates matter and produces prodigious amounts of energy. No other present day theory can do that! Now to explain my original theory that I developed in high school more than 45 years ago, It's really very simple, yet elegant in its own way. It's a variation of the structure of the elements, with the layered orbits just like the elements with the only difference being that alternate orbital layers had differing spins and therefore differing charges. It was the only way I could account for the size difference of the neutrino, $1 \cdot 10^{-9th}$ the size of an electron to the size of a proton, which is 1839 times the weight of an electron, while keeping the electron and proton equal but opposite in charge.

This also explains why matter and anti-matter annihilate each other when they come in contact and produces a prodigious amount of energy in the process. The corresponding different layers of matter and anti-matter cancel out each others spin, thereby releasing energy and annihilating both in the process. Simple! Elegant! It also explains why if enough energy is pumped into ordinary matter, anti-matter can be ejected. Which is exactly what happens during nuclear decay, certain types of photon excitation, X-ray bursts etc. In current theory, matter and anti-matter annihilated each other just after the Big Bang, and only a small out numbering of anti-matter by matter made this Universe possible, then it would stand to reason that there is no anti-matter left in the Universe and there is no way you could produce any. Since this is not so, their assumptions and concepts are misleading and incomplete.

As you can see this theory accounts for spin and charge, something no other theory does.

At present there are No Quantum constructs of the Atom, except the old Bohr construct, which is inadequate to explain either spin or charge. Presently there is no explanation for charge and no explanation for spin. Largest and most important of all there is no explanation for equal but opposite charges for the electron and the proton considering their size and weight difference. The proton is 1839 times as heavy as an electron.

During the last 45 years I have added to the theory to include light, and why it acts as both particle and wave form, something I don't believe present theories allow for, the structure of the Universe and various other conundrums encountered. Along the way I have also investigated the Violins of the Italian Masters and how they were made. I recently finished this manuscript, and am now finishing my work on the structure of the Universe. 45 years ago I came up with an mathematical equation as to why the speed of light can not be exceeded. This Equation is a variation of Einstein's $E = MC^2$. My equation reads $E = MC^2 / (\sqrt{1-(V^2/C^2)})$ where V is the vehicle or masses speed in cm/sec. A related equation for time dilation would read $td = 1 / (\sqrt{1-(V^2/C^2)})$. Although I had heard of the Lorentz transformations for years, until just recently, I had never seen it. Guess where you can find it. That's right, just to the right of the td = sign / as the divisor and in my version of Einstein's equation again as the divisor.

Einstein's theories deal mainly with relativity, the physics and the tensor calculus dealing with the curvature of space-time, this paper will deal with the consequences of the curvature of space-time and the actual structure of the atoms themselves and the Quantum mechanics involved, something I do not believe anyone has ever dealt with in the fashion I will deal with them. I will deal with them in a manner similar to Einstein, as a series of thought experiments. It has occurred to me, during the writing of this paper that my construct of the atom of 45 years ago dealt with the structure of the atom in terms of quantum mechanics for probably the first time, and not in the terms of Newtonian physics as everyone else has tried to construct the atom.

My construct therefore, is a totally different approach to the subject and does not disregard Newtonian physics; it just realizes that Newtonian physics doesn't work at the atomic and sub-atomic levels. The reason is simple; Newtonian Physics deals with gravity, mass and the speed of light and their effects. Quantum Mechanics deals with the strong and weak inter-atomic forces, charge (spin) and shell rotation mechanics and their effects. As you can see they (the two types of Physics) have no common ground on which to write a set of unified field equations. IF the two Physics had even one point of commonality then there might be a chance of writing some type of unified field equations. IT seems the only commonality they have is reality. Which is So tantalizing, as to lead a person to believe there might be a possibility of a unified field theory. Unfortunately, I believe a new type of Mathematics may need to be invented and is going to be needed to connect the two Physics. Remember it's a matter of scale, Quantum Mechanics works in the realm of the very, very small ($1 \cdot 10^{-34}$ cm.) and Newtonian Physics only works only on the scale of the Very, Very LARGE.

Newtonian Physics only works on the scale of the apple and the Earth and larger. A 9 oz. apple dropping 9 feet in 1/2 sec. generates approximately 35,000 ergs. In Quantum Mechanics you are dealing with forces in the neighborhood of $1 \cdot 10^{-9}$ ergs, a light beam generates that amount of force. That's a difference of 13 orders of magnitude. To put things in perspective that's the difference in magnitude of one ounce and 312.5 million tons. It just occurred to me during the proof reading of this section there just might be a connection. It has to do with angular momentum in both relativistic mechanics and quantum mechanics.

First of all, in Relativistic Mechanics, Angular Momentum is ALWAYS present in point particles. This is something Present Angular momentum theories do not allow for these particles to have. In fact you will find there are three separate angular momentums in my construct, the point (Quanta) is spinning of its own accord in either a progressive or regressive mode, it is spinning in a polar orbit around a point in a progressive or a regressive mode, and the whole system is spinning either in a progressive or regressive mode. The proof of this is that if it was a dipole, the rotation rate of the points would degenerate over time and eventually stop. Since this hasn't happened, the rotation MUST be a polar one. Remember there is not just one point, but thousands of points, all in lock step so to speak, spinning in the same direction in the same orbit, with alternate orbits spinning in opposite directions. So therefore, the mathematics for Angular momentum in relativistic mechanics is incomplete.

Secondly, The Mathematics for Quantum Mechanics is also incomplete for the same reasons plus point particles ALWAYS have charge and spin, mass and force are not invariants as present theory would have you believe, they are interrelated co-variants. In Quantum Mechanics you are dealing with as many as 8 inter-related co-variants of equal or nearly equal magnitude all at once. How can you predict anything at that level? The obvious answer is you can't. This is where Heisenberg's uncertainty principal comes into play. Einstein did not understand this and proposed a thought experiment along the lines of two particles hitting each other and rebounding. If you change or tweak one particle does the other automatically respond, no matter how far away it is? According to the math it should, and this caused Einstein and his associates all sorts of problems. They never questioned the math, however, that was their blind spot. The math says the other particle should respond, which Einstein says caused problems with a term he called causality. The solution is really simple. If the Mathematics are correct, then a baseball pitched by a mechanical pitching machine should be pitched to the exact same spot every time. It isn't. Even though the weight of the baseball and the forces applied are the same every time, the inter-related co-variants in this case are some of the same co-variants as the particle: speed, direction of spin, speed of spin, trajectory, direction of travel and time.

My construct would also react the in the same manner as the Stern-Gerlach experiment, but for different reasons. The Classic model of a particle is misleading and incomplete. It is not a dipole or act as a dipole. That structure works in Newtonian physics where the forces involved differ by several tens of orders of magnitude from each other, such as gas planets, suns, moons, satellites, etc. where there is no charge, and relatively slow spin. It does not work in Quantum mechanics where the forces involved are so close to being equals that Heisenberg's uncertainty principal comes into play as a major player. It also does not work where the forces of spin and charge are equal to the other forces involved, and in some cases outweigh the other forces involved. It also does not work because some of the orbits of the quanta would have to exceed the speed of light, which is not possible. I figure the minimum number of quanta per shell or orbit to be 36 (6x6) 6 polar orbits at the same level with 6 in each orbit to a maximum number to be in the neighborhood of 2436. Specifically in the Stern-Gerlach experiment, it was the spin of the entire system that was being affected by the electro-magnetic field. They do have 1/2 spin values as they concluded, but not for the reasons they thought. The reason they got two different spin directions is the particle is constructed of energy with two different directions of spin. I will explain in further detail later. (see further explanation)

The Stern-Gerlach experiment passed a stream of particles through a inhomogeneous magnetic field, with the particles being deflected either up or down by a specific amount, thereby proving the orbits were not random. I am still investigating and trying to find the angle of deflection. I think the precise angle of deflection is important, More on this later also. (see further explanation)

My theory, or more correctly my construct, goes even deeper into the very Structure of the Universe itself, Why its structured the way it is, & Why the Universe couldn't be Structured differently. The Basic Structure is inevitable, suns forming, galaxies forming, Even other Universes. Life is inevitable, as is intelligent life. The Basic Laws of the Universe and Physics Guarantee it, in this Universe as well as in all other possible Universes.

Forty five years ago I was convinced the shape of the Universe could be expressed as an equation or series of equations. I had no idea of the type of equations or series of equations or even the shape of the Universe at that time, more on this later. Parts of the clues have already been provided by greater men than myself. My contribution is that it seems no one else has ever seen the FULL implications of their own work. I didn't put the picture on the puzzle pieces, others did that, I just put the pieces together. Physicists keep trying to picture the universe as a 2, -3 dimensional construct which is an impossibility because it's a 13-16 dimensional construct, More on that later also. The Theory is simple, yet extremely elegant, covers all the bases and doesn't go out on a limb reaching for explanations when an easier explanation is to be found. When you are considering the structure of Everything we must keep Ocam's Razor in mind. The simplest explanation is most often the correct one.

Chapter 2

Others ideas and their unseen implication

The following are a number of examples of not realizing the full implications of what they (the original proponents of the ideas) proposed ($E = MC^2$) 1. $E/M = C^2$, we are all familiar with it, but what does it really mean, and what does it imply? Simply put, it means that when mass is converted to energy it produces a prodigious amount of energy in the process. Energy in ergs = mass, in grams, times the speed of light in centimeters per second squared. An erg is the amount of energy required to move 2 grams 1 centimeter in 1 second. What is never explained is why or how this happens. Nobody has ever explained why C^2? If you have C numbers of Quanta traveling at speed C then you have C^2.

Suppose as a thought experiment, with a little math thrown in for good measure that we take 1 gram of hydrogen (matter) and 1 gram of anti-matter hydrogen and stand back before we combine them. 2 grams of hydrogen has $6.02 \cdot 10^{23}$ molecules of hydrogen, H^2, (Avogadro's #). According to Einstein's Equation, the energy in ergs produced should equal Avogadro's Number. IT DOESN'T! If Einstein is correct then Avogadro's number divided by the ergs produced should equal 1. What we actually have left is the Number of molecules destroyed per erg of energy produced, approx. 668.884.

This number is within the range of my calculated maximum number of Quanta per shell. I figured the number had to be a minimum of 36 or very close to a multiple thereof. Why the numbers 36 and 2436? $2^2 \cdot 3^2 = 36$, $[(3^2)^3 + (3^3) + 2]3 = 2436$. How did I arrive at the number 2436? C in cm. per. sec. /668.884 /1840 = 2436. In the book "Life Decoded" by Marek Lassota it mentions that there are 12 electron quanta energy levels. This would fit. 2436/12 = 203 therefore there would be 2436 quanta per orbit, 12 shells with 203 quanta /shell or ring. There could also be 203 quanta in each orbit moving in 12 directions at the speed of light. Either explanation could describe reality. That type of structure would account for many of the photo emission responses in various elements, chemicals and compounds.

IT ALSO IMPLIES THAT MASS IS NOTHING MORE THAN ENERGY MOVING AT A TREMENDOUS VELOCITY IN A NUMBER OF DIRECTIONS, JUST NOT IN A STRAIGHT LINE. IT'S REALLY SIMPLE!

Mass itself is nothing more than an EFFECT of energy moving at prodigious speeds in at least 14 directions at once. So therefore changing mass back into energy is merely changing form in accordance with Newton's second law. Nothing is lost EVER! I will write more on the full implications of this later. Newton just never went far enough in his statement that energy is not EVER lost, but merely changes form. The strong inter-atomic force, the weak inter-atomic force, are both effects of energy, different wave forms so to speak, each at a different frequency and sphere of influence. The shortest wave form (the strong inter-atomic force) having the greatest influence at short distances and the longest wave form (Gravity) at the longest distances. You could use Einstein's tensor Calculus Equations to compute the range of influence of these forces. No other theory treats these as affects; they are treated as primary forces of nature. Gravity is therefore both an effect of an effect and a wave form although an extremely long one. I've never seen any of these concepts proposed anywhere else.

Another example of authors of various concepts not realizing the FULL implications of their work is that " Space-Time is Curved" This is usually thought of as only occurring around moons, planets, stars, etc. What about the larger implications? Einstein himself didn't see the full implications of this and his tensor Calculus equations.

This implies that the universe itself is enclosed in curved Space-Time, implying that the Universe is an event Horizon, though only on a different scale than a black hole, an event horizon WHERE NOTHING CAN ESCAPE, EVER!

An Event Horizon where Energy and Mass can never be lost only change form
FOREVER!

There is no beginning; there is no end, just a continuous series of big Bangs and Big Crunches endlessly. So maybe both the Buddhists and Newton were right in ways they never imagined.

(Refutation) A third example is that the Universe is only supposed to be about 15.7 billion years old, but the Universe is supposed to be much larger than 31.4 billion light years across. How can this be if nothing can travel faster than the speed of light? So what we are seeing is either the Universe along the curvature of space-time many times or the speed of light can be exceeded or possibly both. An untenable position for physicists if I ever heard one! According to present theories at the age of 200 million years the Universe was supposed to be 60 billion light years across, which is impossible because that's more than 300 times the speed of light. Astronomers even argue over the amount of time before protons and neutrons were formed. Some of them argue that it could have been as long as 380,000 years to as little as 3 minutes before neutrons and protons formed.

Even if it was 380,000 years there is no way it could have expanded to that size in that amount of REAL time. Remember the Universe expanded from a point singularity no bigger than a marble. You figure it out! If protons and neutrons formed 3 minutes after the Big Bang, they, the protons and neutrons, would have had to slow down to the speed of light as soon as they formed. You still can't get around matter traveling faster than the speed of light and the Universe obviously could not have expanded to the proposed 60 billion light years size in 199.62 million years. That's an expansion rate of nearly 158,000 light years per year for 380,000 years. To put it in perspective that's 432.87671 light years per day, or slightly over 18 light years per hour, or .3 light years per minute. THAT'S FOR 380,000 years. 3 minutes is even more absurd. Even over the longest period of time the speed needed would be over 300 times the speed of light for 200 million years which is totally absurd! If we are referring to REAL time, time at speeds considerably slower than the speed of light and relative to a single point, then the above statements are patently absurd. However, if we are referring to relative speeds near, at or above the speed of light, all of the above observations go out the window as I will explain with an example of a near the speed of light object and the time dilation's involved. (See the first paragraph on page 24 of this paper.)

MATTER, THAT MEANS PROTONS AND NEUTRONS, AND ANYTHING ELSE, CAN NOT EXCEED THE SPEED OF LIGHT!

Again I think astronomers need to check their math and rethink these assumptions, theories and concepts! The amount of energy available at the big Bang would have made faster than light speed possible in a linear direction since in all probability the amount of heat energy would have precluded the formation of matter, thereby bypassing the speed limitation according to Einstein's Equation, $E = MC^2$. With no mass in the equation, the speed limit is nearly infinite. Because of the time dilation effect the size of the Universe is indeterminate up to that point, that point being the formation of matter. Remember those times and speeds would not have been real time but relative time.

I believe I have also figured out the minimum mass of the big Bang. That mass is approximately $6.4 \cdot 10^{41}$ solar masses. The solar mass I am referring to is the suns original mass, approximately $2 \cdot 10^{40}$ grams making the mass of the Big Bang event horizon a Minimum of $1.28 \cdot 10^{75}$ metric tons to a maximum of $3.2 \cdot 10^{95}$ metric tons. I've never seen even an approximate mass or size of the Big Bang singularity before. I will explain how I arrived at these figures later. This was arrived at by two different methods, more on the two methods used later. (see further explanation)

In Effect the Big Bang was nothing more than a Super, Super Nova occurring in an event horizon. I've never seen that idea expressed anywhere before either, More on the full ramifications of these ideas later. On proofreading this section I have realized an alternative to this idea which I will present later in this paper, A naked singularity. A fourth example of Astronomers and physicists not realizing the full implications and ramifications of their findings is that the expansion rate of the Universe is increasing. There are three possible explanations for these findings. The first is that the farthest galaxies only appear to be speeding up. Remember these galaxies are millions to hundreds of millions even billions of light years away. Unless you account for the differences in expansion rates between then and now the furthest galaxies will appear to be moving away faster than the closer galaxies. In truth there is only one way to find out if these galaxies are actually increasing their speeds. That way is to graph the speeds of galaxy's expansion at increasing distances and if the line formed (a) is a straight one, then the galaxies are not increasing their speeds and a uniform rate of expansion is in effect. A concave curved graph line (b) would indicate a uniform decrease of speed from a speed near that of light. A convex curved graph line (c) would indicate a uniform decrease of speed from a much lower limited initial speed. A graph line like (d) would indicate a speeding up of the expansion rate. See the graph below.

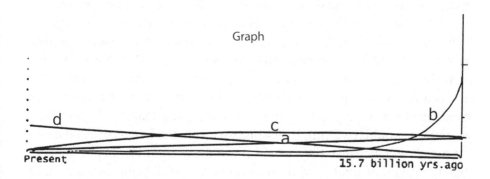

The only way a uniform increase of speed could occur is with a continuous input of energy from the point of the Big Bang. There is no indication of that having occurred or is presently occurring. A jagged graph would indicate a non-uniform input of energy from outside of the Universe or within the Universe. I do expect to see that happening. The reason is simple. New stars are being formed all of the time from the ashes of old stars. New galaxies are being formed from the ashes of the old galaxies and every nova and super nova will disrupt the positions and speeds of the galaxies and stars within them. This will lead to a jagged graph as different stars and galaxies in different parts of the Universe will contribute differing factors that will distort the shape of the original expansion and original shape of the Universe. It would be interesting to see the shape of the graph without the resulting spikes that will distort the shape of the graph or even if there would be enough points left to graph.

I would not expect to see a straight line graph indicating a uniform rate of expansion or a concave curved line graph indicating a uniform slowdown. Since first speculating on this I have found out that it seems I was possibly incorrect and that a uniform speed of 160,000 mph./ 3.2 million years has been verified. (Oct.1999, National Geographic)

That is a uniform rate of expansion and would seem to indicate an open ended Universe, although I don't believe all of the needed information is in yet. I fully expect that when all the information is in it will show a slowly decreasing rate of expansion. Since writing the above paragraph it has occurred to me that if you divided the supposed age of the Universe (15.7 billion years) by 3.2 million years you would get a figure of 4,906.625. IF you multiply that figure (4,906.625) by 160,000 mph. you get a figure of 737,540,000 mph. now divide that by 3,600 (the number of seconds in an hour) and you get a rate of recession that is 204,872.22 miles per second, which is faster than the speed of light (186,234.2 miles per second), 1.1 times the speed of light, which is impossible because any galaxy further away than 14 billion light years would simply vanish from our view. This hasn't happened. A recession rate of 160,000 mph. /3.2 million years works out as a recession rate of .00002006 inches /per second. For a uniform increase of speed, any galaxy that is further away than 6.4 billion light years away would have simply disappeared from the Universe as far as we are concerned, because they would move away from us at speeds faster than the speed of light relative to us. This has NOT been shown to have happened, so this option is not tenable. A uniform expansion rate would have shown similar results at the 10.4-13.4 billion light year range, as I just showed, again since there is NO indication that this has happened, this option is NOT tenable either. The third option is no more tenable than the first two, as the speed of expansion would have exceeded the speed of light 900 million years ago, before there was life on Earth.

THE UNIVERSE MUST BE A CLOSED EVENT HORIZON!!
IT STARTED OUT AS ONE AND IS STILL AN EVENT HORIZON!!

Since no galaxy or object has been seen to exceed a recession rate of greater than 25% of the speed of light according to red shift, the above figure of 160,000 mph. /3.2 million years is obviously wrong. From the facts that no galaxies are receding at a rate of faster than 25% of the speed of light, and the Universe is approximately 15.7 billion years old, the maximum rate the Universe could be expanding is .00002 inches a second / per second in all directions. The Hubble constant, which is a straight line expansion rate of 1,700 km. / million years works out as an expansion rate of .0000539 mm / second. = .000002 inches /second, 1/10 of the above rate of expansion. This would tend to indicate the rate of expansion is decreasing from an average of .00002 in. / sec to a rate of .000002 in. /sec. in 15.7 billion years. THAT looks like a slowing down of the expansion rate of the Universe. The light from the furthest reaches of the Universe would have left their point of origin nearly a billion years ago, BEFORE there was life on Earth. This information might be used to determine the size of the Universe. ALL of the above observations go out the window of course if we are in a closed event horizon Universe. The why will be explained later in this paper.

Chapter 3

Introduction of Fractals, Mandelbrot sets & Julia sets

It has also occurred to me that acceleration is NOT ABSOLUTE. NOTHING in Nature is Absolute. EVERYTHING is Relative. I will refer to Einstein's relativity as uniform relativity and the new relativity of acceleration as changing relativity, since the relationship in the new relativity is in a constant state of flux. A possible equation to express such time dilation as would accumulate in acceleration would be
Td = $\{a \cdot 1/\sqrt{1-(V^2/C^2)}\}^{2t-1}$ where td = time dilation, a = acceleration in meters per sec,
V = accelerating body's velocity in meters per second, C = speed of light in meters per second.
t = times the process is repeated this is not a power but is cumulative.

Einstein dealt with this subject in his Nov.1915, Thursday lectures before the Prussian Academy of Sciences with his famous tensor Calculus equation. $R_{uv} - 1/2 g_{uv} R = 8\pi T_{uv}$

Even Absolute zero is not Absolute, Again it is relative. The motion of all of particles and matter is at minimum, there is no extraneous motion. (see further explanation)

This is the only time matter does not appear to act like strings, More on this concept later. (see further explanation)

The shape of the evolving Universe from the Big Bang to present day can be expressed as an evolving Mandelbrot or Julia set. Remember both Mandelbrot and Julia sets start off with x = 0. What is the three dimensional shape of a 0, obviously a sphere and a Mandelbrot set is formed within the limits of an egg shape, if you feed a series of numbers into the equation you get a series of shapes evolving into a pattern that can be found almost anywhere in nature.

Mandelbrot sets can be expressed as $f(x) = x - u$ where u is a constant. Julia sets have been expressed as $f(x) = x^2 - u$.

However both these figures are possibly wrong or misleading as $\pm (x + u)(x - u) = x^2 - u^2$ not $x^2 - u$.

Remember these can be both graphed points and iterations that are sequential, repeating and form patterns that are fractals resembling most of the forms found in Nature. Strictly speaking Mandelbrot sets are $f(x) = x+u$ or $x-u$, while Julia sets are $+$ or $-$ $f(x) = (x +u)(x-u)$.for graphed points and are $f(x) = x^2-u$ for iterations.

Remember Mandelbrot and Julia sets are the only equations that express infinite patterns and sequences in a finite space. However fractals, both Julia sets and Mandelbrot sets, are only applicable to two dimensional space and not a three dimensional or four dimensional reality. After much thought I have come up with the following set of equations; for graphed points in the evolving set.

x, y, z = $f(t)$, $f(x) = +$ or $-(x + u).(x - u)$, $f(y) = +$ or $-(y + u).(y - u)$, $f(z) = (z + u).(z - u)$ $f(x) = f(y) = f(z)$, where x, y, & z are the x, y,& z axis and t is time and f is a function. u is a constant, probably a primary Eigenmode of hydrogen, possibly .028 cm. Why .028 cm? $_3\sqrt{3} - \sqrt{2} = .028$. Use the other form for iterations. $f(x) = x^2 - u$ for y and z. These equations give only a part of the evolving shape of the Universe as they do not include the Eigenmodes, Primary, Secondary etc. modes, plus they do not contain the re-enforcement and cancellation frequencies and modes. IF you feed the figures for x sequentially increasing from 0.0 to 2.000 you get part of the evolving shape of the Universe.

Again the figure for u is probably a function of hydrogen's primary mode of vibration (Eigenmode) or energy peak. I suspect .0028 cm. This is for graphed points. Then there is the matter of up, down & strange particles. Up, down and strange particles are I expect; nothing more than merely stable decay products, not elementary particles. If electrons, protons, etc. can only break down in certain patterns and still be stable, so it is with these particles. This doesn't mean these are the basic building blocks.

It just means these are stable by products of nuclear decay. The basic building blocks are much smaller, NOTHING MORE THAN SPINNING ENERGY ITSELF. Neutrinos were unknown when I first developed this theory, more actually a construct, they were only theorized.

They are still a dilemma for physicists; they don't have any idea as to the structure or any idea about their characteristics. My theory explains both two possible structures and their characteristics. Present theory also does not account for the difference in mass in protons and electrons but their having equal but opposite charges. Present theories also do not account for spin, Except thru a complicated manipulation of particles that have never been found in nature. Mine does in a simple, yet elegant manner.

Chapter 4

Definition of Quanta, The Paradox of light & consequences

It has also occurred to me that light itself is a paradox. While it seemingly can't travel faster than light it must travel several times faster than light just to be light. Let me explain. First of all, light is both a particle and a wave form, they don't just act like particles and wave forms. They are both. (The English language is inadequate.) Einstein also dealt with this, but never saw the full implications and only approximated it with his equations. As a particle light vibrates, or more correctly spins [1·10^{56} rps. for a light particle] within its own event horizon, energy packet, shell [take your pick] (Again the English language is inadequate) in a progressive and regressive orbit at the speed of light.) Light moves in a wave form at greater than the speed of light [I'll explain later] and travels in a linear direction at the speed of light. Getting back to the wave form explanation, Light, I suspect, is a 3 dimensional wave form, not a two dimensional wave form, as it is usually depicted. That is why we have polarized light. Right handed and left handed. That means it [light] travels in a spiral fashion, not a straight line as is usually depicted, but a corkscrew fashion as it were, The higher the frequency the tighter the corkscrew. This also means the tighter the corkscrew [the higher the frequency] the further the particle, energy bundle, photon, quantum, take your pick, travels according to the formula,

$TL = \pi$ D Fl N Fs /Cos. Angle. Where TL = total length traveled in one second,

$D = (d^1 + d^2 + d^3 + \ldots)$ diameter of wave form in meters, miles, km., $\pi = 3.1415926$, Fl = frequency length in meters, N = number of Rotations per frequency length, Cos. Angle = Cosine of the Angle

As you can see light must actually travel many times faster than the speed of light, it just can't travel in a linear direction faster than the speed of light. This may seem to be splitting hairs but it explains where the energy goes as more energy is applied to light and/or another object including matter. Einstein himself had problems with his famous equation: $E = MC^2$ and postulated 2 exceptions that are not usually presented in most literature.

The first is the better known because it has to do with black holes and is sometimes referred to as the spaghetti effect. It is usually explained with matter as an example but it would cause the same effect on light, that is it would cause the light waves, particles to become linear, pulling them straight as it were, thereby causing a seeming "warp" in space-time. Seeming to make time stand still, while it pulls the light wave, particles linear (again the English language is inadequate) apart in a linear direction. The other exception, which I have never seen so stated but is obvious, would be the flip side of the above situation, Squirting light or matter out of a small self contained event horizon, thereby seeming to exceed by many times the speed of light. Theoretically possible, although the energy requirements would be prodigious, controlled fusion type stuff.

I have just read about a similar example of this in Paul Davies book "About Time" where on page 82 he mentions a neutron outburst from Cygnus X-3 situated about 35,000 light years away. Since a single neutron has a stable half-life of only about 15 minutes at rest, the time dilation effect is prodigious. {That's a time dilation of nearly 12.250 billion to one}. This speed is .0096 of an inch a second less than the speed of light.) That's Squirting a pumpkin seed out of its skin while bringing the skin along with it so to speak. Since light is both a wave form and a particle I will refer to it as a photon from this point on, It could also be referred to as a quanta, both would be correct. You could correctly say it is a particle traveling on or as a wave form and a wave form spinning so fast it gains mass and acts as a particle. Both would be correct.

$(E = MC^2)$ $[E/M = C^2]$.

Each photon (quanta) spins on its axis of rotation (x-axis) at the speed of light. (0) {$3 \cdot 10^{56th}$ rotations per second}. Each photon (quanta) also spins in a progressive or regressive football shaped orbit (1) (x-axis) that would make the photon (quanta) seem to vibrate. It also spins around a center of rotation [approximately 1.10-34 cm.] {Planck's Constant}(2). The vector of these two forces is a complex wave form that seemingly travels faster than the speed of light (3).

This entire structure also spins around another axis of rotation (y-axis) at the speed of light (4) {300,000,000 meters per second) and also spins in a football shaped orbit (y-axis) (5). Again this spins around a center of rotation that seemingly travels faster than light (6). This again produces a vector (7) that seemingly travels further than 300,000 km per second. If this isn't enough to confuse anyone, It (the photon) (quanta) also spins in the Z-axis (8) around its axis of rotation. Like 1 & 5 (9) also spins in a football shaped orbit only in the z-axis, 2 & 6 are similar to 10, only 10 spins also in the z-axis.11 is similar to 3&7 only in the z-axis. All of this is moving in three dimensions [12(x-axis), 13(y-axis), 14(z-axis)] simultaneously as a wave form frequency traveling in a straight line at the speed of light.(15) All of this adds up to a minimum of travel of more than 12 times the distance that light should travel in a straight line in a second. The more energy added to light the more 1,2,3,5,6,7,9,10,11,& 15 change. Actually it's more complicated than that. 12,13,14 combine to travel faster than light (or I should say more correctly the photon travels in one second a distance that is greater than 300,000 km., which is the speed of light in one second.) This may seem splitting hairs but it is very important, because although it (the photon) travels further than light should travel in one second it does not travel in a straight line faster than the speed of light.(It probably has something to do with the time dilation effect and mass accumulation effect.)

To determine the distance the photon (quanta) traveled we must use the formula
Tl = π DFlNFs /cosine of the angle:
Where Tl = total length traveled in one second, π = 3.1415926 . . . ,
D = (d^1+ d^2+ d^3+ . . .) Diameters in meters, miles, km, etc., Fl = frequency length in meters
N = number of rotations per frequency length. Fs = Frequencies per second.

Think of a string wound around a football. The football represents the wave form of light or any radiation. Let's take several wave lengths as examples. The first has a frequency length of one hour, the second a wave length frequency of one second, the third a wave length frequency of 1/3600 of a second. The first travels only marginally further than

300,000km. (Approximately 303,000km.), the second travels nearly $\pi/2$ times the distance 300,000 km x 1.5707913 = 471,237 km.

These are only the figures for vectors 12, 13, & 14 for the first two examples. Vectors 0, 4, 8 spin at the speed of light. The Vectors 2, 6 & 10 travel distances greater than 300,000 km. per second, as probably do vectors 1, 5, & 9. This is where part of the energy goes when more energy is added to the system. The distances for vectors 0-11 have not been included.

$E/M = C^2$: Since C & C^2 are constants, any energy added to a system [whether it is light or matter] will result in a corresponding increase in mass. Conversely if C is a constant any loss of energy in the system will result in a corresponding loss of mass in the system. The implications of these ideas are huge! This is why light acts like both a wave and a particle. It's spinning so fast in so many directions at the same time that it gains mass in the process. This is how and why light can act as both a particle and a wave form. This also explains why matter acts as strings, its spinning so fast, in so many directions, in such a small space that it appears to act like strings.

This is why matter appears solid yet it is 99.999,999,999,999,999,99 9,999,999,999,999,999,999,999,999,999,999,999,999,999,999,999, 999,999,999,999,999,999,999,999,999,999,999, % nothing.

(Strings) When string theory is referred to by mathematicians and physicists are they sure strings are real or are they merely the ripples caused the stone thrown in the pond ? There is the distinct possibility that what these people believe are vibrating strings are merely the ripples caused by rapidly spinning particles spinning around incredibly small centers, spinning at such incredibly high speeds that they appear to act like strings. I believe those in physics should concentrate on the stone and not the ripples. These rapidly spinning particles are spinning at speeds in excess of $3·10^{56th}$ rotations per second/per second, Around centers that measure less than $1·10^{-34th}$ meters. If the spin is processional (Right Handed) the charge is positive, if the spin is recessional (Left Handed) the charge is negative. (Single points of energy Spinning like a top at the speed of light.) The speed of these particles is what makes matter appear solid. Most people have a hard time with that concept. Even some physicists have a hard time with that concept.

If all of the empty space is removed from the Universe, the rest of the Universe could be compressed into a ball about the size of a basketball or even smaller than that, probably as small as a marble. The stone (energy {photon} [quanta]) is what constitutes the basic building blocks of the Universe, Our Universe {3 Dimensions}. Those spinning points of energy are the building blocks of the quanta and all of nature. These points of energy do not spin in perfect circular orbits around their centers, They spin in a () football shaped polar orbital pattern around their centers of motion. The more energy added, the greater the pattern is distorted. This is why the particles seem to vibrate, and act like strings and also where much of the added energy goes. It is also why the speed of light cannot be exceeded except in special circumstances.

There are no parallel Universes or parallel dimensions, Or higher Dimensions other than more Universes (3 Dimensions). There are however the lower Dimensions just mentioned, those at the sub-atomic levels [3 Dimensions] and those at the atomic level {3 Dimensions} + Time (1 Dimension). That makes 13 Dimensions to possibly as many as 16 Dimensions, exactly the range of present day string theory. How can TIME be only 1 Dimension. I believe Time flows from + to),),),)to >. I also believe the rate of Time flow or passage is different at each level. I believe the rate of flow at the sub-atomic level is $1 \cdot 10^{-9}$ sec., at the atomic level $1 \cdot 10^{-3}$ sec., at the solar level it would be $1 \cdot 10^{3}$ sec., (approximately 16 2/3 light minutes) which is approximately 1 Astronomical unit, at the galactic level its approx. 3 light years $1 \cdot 10^{9}$ sec. and on the Universal level $3 \cdot 10^{18}$ light years. (This is approximately the size of the Universe).

Chapter 5

Time & Entropy, Explanation

Think of Time as an arrow or line moving from + [the big bang] thru) sub-atomic,) atomic,) our Universe,) Other Universes to some point >. Think of Time as slowing down the further it gets from +. Entropy runs in the same direction from +,),),),), to > getting greater as it moves from +,),),),)to some point>. If we are in a closed system, a closed universe with enough mass, then Gravity will cause matter and all the energy to go in the other direction from. < thru (, (, (, (, to +. [In this case the Big Crunch]. No energy is lost; it is merely transformed into another form in accordance to Newtonian Physics and in the Big Crunch can be changed back into anything. The Universe is itself a closed event horizon. As I showed earlier, No energy in any form can escape EVER!!, Thereby conserving Newton's Second law.

Present theory does not allow for this to happen or even confront that possible eventuality. Again Einstein only touched on this in his Thursday presentations during November of 1915. Einstein was sure his theory did not eliminate Newton's laws, he was sure these laws just helped define Newton's laws and refine them. Unfortunately Einstein never made this connection.

The names of the particles have changed so much since I originally came up with these ideas back in 1962; that I had to do considerable research just to recognize those particles I knew so well 45 years ago. The names of bosons, leptons, gluons etc. were unknown back then, but I had already grouped some of the particles as they are now grouped according to mass, putting those with similar masses in the same group regardless of charge or spin.

Chapter 6

2 possible explanations for the neutrino & analogy

I noticed 45 years ago that each group decayed into the types of particles that were similar in mass and only differentiated slightly in spin and charge. The atomic weights of the decay products were unusual in that they were masses with additional fractions of the mass of an electron, Smaller than a proton, but greater than an electron. I remember one as 437.4 or 438.4 times the weight of an electron. This struck me as odd, for electrons were supposedly the smallest stable particle. The only so called particles smaller than an electron, the neutrino, had not been identified at that time, only theorized. This got me to thinking that the neutrino (Which I suspected weighed approximately $1 \cdot 10^{-9}$ the weight of an electron) was itself made of a + right particle and a - left particle. {I named them that for the direction of their spin.}

 I surmised that there was the possibility that the neutrino might even be more complicated than that. The distinct possibility existed that it might consist of multiple layers of spinning packets of energy (quanta). This is a structure similar to an onion with each layer having its own energy level. This made sense for these particles only decayed at certain energy levels and only combined at other energy levels. They were consistent. Unless a similar structure existed as the basis for the form of atoms, there would be no way to pre-determine decay products and energies. Since we can predetermine both decay products and energies such a structure or structures must exist.

 This also accounts for the production of anti-matter by the insertion of high energy particles on matter. Present theory doesn't know how to explain this situation. As a thought experiment, think of an onion with alternating layers of red and white layers, each layer represents a state of matter and stable energy levels. The white layers represent positive matter and the red layers represent negative matter, anti-matter as it were. Each stays in their own layer and do not interact with each other, because each layer is a separate energy level and each layer is stable unto itself.

Now suppose we inject a sharp needle (a high energy beam or particle) penetrating the layers of the onion (atom) and removing several parts (kicking a part of the atom) of both the red and white layers (matter and anti-matter) out of the onion (atom) and out of their respective stable layers. Then the part of the onion (atom) red and white (matter and anti-matter) are forced to act as is their nature outside of the onion (atom).

Inside the onion (atom) the respective layers are stable and cause no harm, but outside of the onion (atom) they can cause a big stink in more ways than one.

Chapter 7

Possible shapes of the Universe + consequences

There are presently two proposed shapes of the Universe. The First is the sphere of a closed, event horizon, Universe. The other is a Saddle shaped open ended ever expanding Universe, This later form is a two dimensional construct and is not relevant in the 3 or 4 dimensional construct that is our perceived reality. A saddle form only applies in a 2 dimensional construct, as a flat plain also only applies on a 2 dimensional construct. A more accurate 4 dimensional construct for Euclidean space would be a sphere and a 4 dimensional visualization for a saddle shape would be a square beignet with a hole in the middle. I will show how the last shape is not the shape of an ever expanding Universe except under special circumstances and these special circumstances could just as likely produce a closed event horizon Saddle shaped Universe. At the present time most people are familiar with the two forms of Geometry: Euclidean and Boolean. Boolean geometry says that parallel lines meet at some point.

Euclidean geometry of course says parallel lines never meet and stay equidistant forever.

I have thought about geometry on a Universal scale and have come up with two possible geometries that I have just seen before proposed in print. A third possible geometry, only applicable on a Universal scale, is that Parallel lines only appear parallel, but in fact they diverge to infinity. This is only possible in an ever expanding open ended unimpeded Universe. Such a Universe is still a sphere, not a saddle. It's only a saddle as a two dimensional construct starting from a point 0 at rest, not a 4 dimensional construct starting at a point with a large input of energy. This is called a hyperbolic Universe.

The Fourth possible geometry is only possible within an event horizon the size of the Universe. This geometry is the result of the effect of the curvature of space-time. It also appears on a smaller scale in electromagnetism around magnets and around the earth's electromagnetic field, the lines are both curved and parallel. Parallel lines only appear parallel on a small scale, while they truly aren't just parallel. They are both curved and parallel. This is Boolean geometry applied on a Universal scale. This would be nearly impossible to prove as the base line for any such experiment is likely to be either too short to measure any difference or will be distorted by the curvature of space time itself. Intelligent yeast would have the same problem trying to prove if the earth was round or flat. The scale we would both be trying to measure is so vast, that any possible measurements either of us could make are so overwhelmed by the margin for error as to be useless.

Einstein hinted at this with his approach to relativity by methods of non-Euclidean Geometry and Ricci tensor Calculus, but again he never realized the full implications of his ideas or calculations. Both of these Universes are spheres, when in their initial stages, when they are unimpeded by another Universe. The only difference shows up later. If enough mass is present, then the Universe is an event horizon Universe and falls back to a Big Crunch. If it doesn't contain enough mass, then that Universe will expand forever unless impeded. The only way a Saddle shaped Universe could be formed is to have this Universe (our Universe) be only one of Many Universes. We have no evidence of this, however, if our Universe is a closed event horizon Universe, there would be no evidence of other Universe's because of the curious nature of event horizons. I will write more on the curious nature of Event horizons later in this work. The only way a Saddle shaped Universe could be formed is for the Big Bang of Our Universe to have occurred near another Universe, close enough to impede the expansion of Our Universe, thereby distorting the shape of our expanding Universe. It would make no difference if our expanding Universe is a closed event horizon Universe or a open ended ever expanding Universe. The initial shapes would be the same. How close the impeding Universe was at the time of the Big Bang, The rate of expansion of our Universe, the movement of the other Universe and its size would, of course determine the amount of distortion of our Universe and our Universe's eventual shape. In this scenario it would be possible for one Universe to envelop and consume another Universe, and inhabitants of neither Universe would know the other existed until just before one Universe consumed the other Universe.

It has occurred to me that whether our Universe will be an open ended, ever expanding Universe or a closed event horizon Universe is a mute point at this time. The Universe is still too young to know its eventual outcome; the Universe is only 15.7 as it were. It's still a teenager. Much too young to know what she will look like in her old age, or to predict her shape in her old age. The shape of the Universe is still one of an event horizon Universe and will be for approximately another 20 billion years or so. By that time it will be irrelevant to anyone or anything living on this planet by approximately 15 billion years. Within the next 5 billion years, the Andromeda Galaxy, which is moving on a collision course with the Milky Way [Our Galaxy] at a speed of 250,000 miles an hour, will collide with our galaxy and whether our solar system survives such a collision will be strictly a matter of luck. That Event, as tragic as it may be for future inhabitants of our solar system, or our galaxy, will not change the basic structure of our Universe. Even if that doesn't happen, our sun will have started to expand, as it will have burned up almost all of the hydrogen in its nuclear core and will start burning helium, Which will make it too hot for any life on earth.

Chapter 8

Problems determining if we are in a closed event horizon

Even in 5 billion years the shape of our Universe will be that of a closed event horizon Universe and the geometry and physics of a closed event horizon Universe is markedly different from that of an open ended ever expanding Universe. How can this be you ask? Light and matter will act differently in closed event horizon Universe than in an open ended ever expanding Universe. How you ask? Since Light and Matter can NOT escape a closed event horizon EVER, the path light and matter will take once the boundaries of the event horizon are reached will take a curved path and could theoretically return to the same point from which they originated, although millions or possibly billions of years later. Since light and matter would travel in a curved line and not a straight line, there would be no direct way to determine the size or amount of matter in a closed event horizon Universe. More on a possible means of determining both size and shape of the present day Universe will be presented later in this paper. There would be no direct way to determine if the Universe was 2 billion, 20 billion, 200 billion or 2,000 billion light years across. Einstein hinted at this too in his tensor calculus equations but never seemed to understand the full implications of what he proposed. General Relativity does not say that the Universe is saddle shaped it only suggests that shape might be one of the possible shapes. Nobody else has understood the full implications of his work either, until now. Since there is no outside reference point, there is no direct way to determine the size or shape of the Universe [more on this later.] and an outside reference point would be useless to someone inside a closed event horizon. A real paradox if I ever heard one. Such is the bizarre nature of event horizons.

    There is other information that seems to confirm that we are in a closed event horizon Universe, and that is that no matter which direction you look, all you can see are quasars or their equivalent, which would make one tend to believe that we are indeed in an event horizon Universe. If we were in an open ended, ever expanding Universe you would be able to look in almost any direction and see open areas, and possibly other Universes. Since no such areas exist, we are almost certainly in an event horizon Universe as I stated earlier. It would be similar to intelligent yeast trying to figure out the size and shape of the 100,000 # self-rising sweet currant bread that had 10,000 # of currants. [The 10,000# of raisins are the approximate 10% of the Universe we can see, and the 90,000 # of self rising sweet currant bread, the portion of the Universe we can't see.] We are in the same predicament. As an intelligent yeast, we are trying to describe the size, shape, and eventual outcome of that 100,000 # self rising sweet currant bread. Most of the ideas as to its eventual outcome (size and shape) are only as they say, pardon the pun, half-baked. Maybe this will get a rise out of some astronomers and then again maybe not. Now to explain Primary and Secondary Frequencies, Primary and Secondary Harmonics, Eigenmodes, Nodes, Anti-Nodes, Etc.

Since we have established that matter is nothing more than vibrating energy; and Eigenmodes are the Primary frequencies of a vibrating body, we and everything in this Universe are the result of the Harmonics of the Big Bang. The Primary, Secondary, Tertiary, Fourths, Fifths, Sevenths, Eighths, etc. are the result of the Big Bang occurring at a naked point singularity, a collapsed event horizon if you will. The weight, size, and density of the point singularity that was the Big Bang could possibly be determined by knowing the exact wave length of the background noise of the Big Bang (It's in the micro-wave range). Since we know that when a neutron star of approximately 1.44 times the mass of the sun (The CHANDRA limit) and above, collapses it produces a internal shock wave of approximately 300 cps. The physical reason for the Chandra limit is this is the point where the electron repulsion balances the force of the collapsing gravity. Any mass above this amount causes the star to collapse even further to the neutron repulsion balance point, Approximately 3 solar masses. The physicists still don't know why this figure (the Chandra limit) is significant, but I think I know. The $_3\sqrt{3} = 1.44224957031$ (the Chandra limit), $(2._3\sqrt{3})^2 = 8$ solar masses which is the mass of a star necessary for it to go nova. The final mass of the remnants of the nova or super nova can not be over approximately 3 solar masses or the star will probably collapse and become a black hole. If the final mass is under the Chandra limit, then it will probably become a white dwarf, If the final mass is over the Chandra limit but under 3 solar masses it will in all likelihood become a neutron star. I believe $_3\sqrt{3}$ has other significance in the iteration of Julia sets, where x is the $_3\sqrt{3}$ and u is the Eigenmode of hydrogen. If this is true, then the Universe is a closed event horizon. (Incidentally one of the values of the Eigenmodes of Hydrogen is .028 cm, which can be obtained by subtracting the square root of 2 from the cube root of 3)

These are not exact limits, because a star with a very high rotational speed can have higher mass than these limits, of course once it slows below the speeds necessary to maintain the balance between gravity and its internal pressure it would evolve into the next stage.

$(2^2 \cdot {}_3\sqrt{3})^2 = 24$ solar masses which is the mass of a star necessary for it to become a super nova, after it's burned all of its nuclear fuel, again the final mass of the remnants of the super nova can not be over approximately 3 solar masses or the star will probably collapse and become a black hole. If the final mass is under the Chandra limit, then it will probably become a white dwarf, If the final mass is over the Chandra limit but under 3 solar masses it will in all likelihood become a neutron star. These again are not exact limits, because a star with a very high rotational speed can have higher mass than these limits, of course once it slows below the speeds necessary to maintain the balance between gravity and its internal pressure it would evolve into the next stage.

$({}_3\sqrt{3})^3 = 3$ solar masses is the minimum mass necessary to become a black hole, this is of course after all nuclear fuel has been burned and the core of the star has collapsed,

$(4^3 \cdot {}_3\sqrt{3})^{18} = 6.4 \cdot 10^{41}$ solar masses. This is the Minimum Mass of Big Bang before the Big Bang took place. To $(5^3 \cdot {}_3\sqrt{3})^{27} = 3.2 \cdot 10^{56}$ solar masses for the maximum mass for the Big Bang. This makes for a minimum of $1.28 \cdot 10^{76}$ metric tons to $6.4 \cdot 10^{96}$ metric tons maximum. Since we also know that the frequency of the wave increases with both an increase of the density and a decrease of the size of the vibrating body, from this information we should be able to extrapolate the size and density of the original point singularity that was the Big Bang.

Chapter 9

Big Bang misnomer, explanation + consequences

Strictly speaking, "The BIG BANG" is a misnomer, what I believe happened was really simple.

The point singularity had a mass, charge and/or spin rate that exceeded the limits of their combined event horizons and therefore became a naked singularity. Above 3 solar masses gravity forms an "event horizon" around the mass, effectively cutting it off from the Universe. If the star also has spin, a second "event horizon" is formed inside the first "event horizon", the mass event horizon forming a secondary spin event horizon inside of it, the greater the spin, the larger the spin event horizon becomes and the smaller the mass event horizon becomes, If The body also Had a charge when it collapsed, that would form a third event horizon causing the other two event horizons to become even smaller. If All three event horizons are the same size and any mass, spin or charge is added, the event horizons simply disappear and the forces holding them together become unleashed. With gravity now acting as a repelling force (electron and neutron repulsion) the Universe was born. No Big Bang, no explosion as such, No annihilation of anti-matter and matter, just a rapid expansion of a very hot accumulation of energy that latter re-organized itself into matter. It's Simple really, and all from a naked singularity. So maybe it should be called "The Big Rejection" or "The Big Kiss Off", or maybe "The Big Push," you get the idea.

The reverberations of that event off of what was left of the event horizon and the remains of its contents created ALL of the above frequencies and the Universe is the result. How you ask? All of the above frequencies and energies reacted and interacted to each other in such ways that they re-enforced each other, canceled each other out or just created white noise and isolated energy. This concept is very important! That white noise can be heard as all that is left of the misnamed Big Bang, and is everywhere present in the Universe, As Penzias and Wilson discovered in 1965. If we are not in a closed event horizon Universe the reverberations from the misnamed Big Bang would have dissipated and disappeared long ago and would no longer be detectable. The so called Seifert galaxies may have black holes at their centers that have gone through this process. That explanation would account for the massive outburst of energy pouring out of these points and the seeming rapid expansion of the cores of these galaxies. NGC1275, Messier 87, NGC5128 are three possible candidates. The original differences were not large, but they were just enough to cause variations throughout the newly expanding Universe. Without these minor variations, which I believe are inevitable, the Universe as we know it would not and could not exist. The Eigenmodes of the primaries, secondary, etc. of the reflections and refraction's from the inside of the singularity that was the Big Bang practically guarantee a Universe similar to ours. The places in the Universe where the energies and frequencies canceled each other out, became voids where nothing exists today.

Places in the Universe where the energies and frequencies re-enforced each other eventually became matter in the very hot newly expanding Universe. This all has to do with fractals and evolving Julia sets in 13-16 dimension space-time. I will deal with these concepts later and it will tie all of these various aspects together. In the Early state of the expanding Universe, probably for the first 30,000 years to as much as 380 thousand years after the Big Bang it was probably too hot for matter to have formed. It was during this time that the greatest expansion of the Universe undoubtedly took place. Again the size and shape of the Universe is possibly indeterminate, as we have no point of reference. If space-time is as truly as curved as I believe it to be, We, and Everything in this Universe are in a closed event horizon, A closed event horizon of truly extremely gigantic proportions. I believe the Big Bang came about because of the collapse of a previous Universe and we are in the beginning of another expansion stage of a never ending cycle of Big Bang expansion to Big Crunch compression. The entire Structure of the Universe depends on the Quantum construct of matter. Understanding what happens inside a black hole depends on your understanding of what's going on in the Quantum construct. Since you the physicists have no Quantum construct that fits all the parameters with out a lot of stretching, guessing, hoping and downright wishful thinking, how can you possibly understand what's going on in the Universe or a black hole?

There are no white holes, no alternative Universes. What's your white hole, Just a naked singularity. After much thought, ruminations and an epiphany or two it has occurred to me that both the physicists and I could be both right and wrong at the same time. If the Physicists are looking at the larger structure, and I'm looking at the smaller structure we could both be correct. What if one stable particle size was 668.8 electrons wt. and had a positive charge, and you had two of them and another particles wt. was 437.4 electrons wt. with a negative charge and 62 electrons wt. of gluons and assorted particles to hold it all together. It could still work for both theories. So much for the relatively small, now for the Large.

Chapter 10

The Evolution of galaxies

I believe all galaxies start life as a spheres, and as their spin increases they become more and more disc shaped as their speed increases (Accretion disc), as their speed increases even further they start forming spirals, the faster the spin, the more spiral arms develop, the more spiral arms develop the more space is compressed ahead of the spiral arms, becoming the nursery for new stars, So much for the size, shape, structure, and outcome of the Universe. I have since thought of a way to determine both the size and shape of the Universe. It has to do with the red shift and the age of the Universe. There is also the possibility the Universe could be in the shape of a donut, both expanding and contracting at the same time. I will write more on both of these later and the consequences of both of these, and methods of determining if they actually exist. In order to determine both of these matters it will be necessary for astronomers to map the Universe in ways it has probably never been mapped before. It will probably take mathematicians to determine where the stars and galaxies are now, not where they were when the light was emitted from them. This may seem trivial, but it is important in that it will reveal the present shape of the Universe, not a jumbled mixed up hodgepodge of star and galaxy positions through out time. That situation would give us no information as to the present or past shape of the Universe.(see further explanation)

Each star galaxy and galaxy group needs to be mapped as to distance, direction, speed and sector of the universe and present position. Only then when we have all the galaxies and stars in the same time frame will we know the actual shape of the Universe. This will probably involve tensor calculus. It will probably also involve the use of several very large virtual reality computers and the use of many computer programmers to input the locations of the stars and galaxies. Again this will probably involve the use of tensor calculus and many many, many years of programmer input and computer time. This is Definitely not a project for the faint of heart or someone looking for Quick results. Now for the donut theory, I know what you're thinking, that I'm going to shoot holes in some other theory. Not so! Just as a thought experiment consider if the Universe were in the shape of a donut and not self rising sweet raisin bread. Then the Universe would be expanding and contracting, enclosed and open ended at the same time such a Universe would seem to have no contradictions but the astronomical observations observed by the inhabitants of such a Universe would be very normal indeed. You would still have Curved space time, in fact all of the observed data would look no different if you were in a donut shaped Universe than they would for A open ended Euclidean Universe, A closed Boolean Universe or a Mr. Donut Universe. The only one who could tell the difference is somebody outside the Universes, because on the inside there would be no way to tell. It has to do with not having reliable reference points. In a Mr. Donut Universe you would get conflicting evidence for both a closed Universe and an open ended Universe, as you would for both a closed Universe and an open ended Universe in each of the other types of Universes. In fact some of the Reimann equations would then be relevant in a Mr. Donut Universe, where they would be only of mathematical interest in either of the other Universes. The proof of the afore said Quantum Construct is that when a neutron star collapses, the structures of the construct quanta actually collapse into pencil shaped structures for the atoms.

This shape is a natural out growth of the added energy introduced into the system because of the stars collapse. Just like the added energy distorted the shape of the atoms and the quantum construct originally and seemed to make them appear as strings, so the additional increased energy and compression from the collapse of the star caused the atomic construct to change form once again under the additional influence of extremely strong magnetic fields. Why does it form needle or pencil shaped atoms when a sphere is the most compact form? It's because it has to. It can not cross magnetic lines of force, it must travel along them. The most efficient shape to do that is a needle or pencil shape. Further proof that my construct is correct is the atom and the quanta can not cross magnetic lines of force proving that both the atom and the quanta are charged and are influenced by magnetic fields and also have spin. Further proof of my Quantum construct is that nobody else has been able to explain why $E = MC^2$ is correct.

They just state it as fact with no explanation as to why or how it is correct. My construct shows it and proves it both at one time and in simple terms. My Quantum construct also satisfies all of the things we know about Quanta, the atom and matter but no other theory can explain, Charge, Spin, Annihilation of matter, Anti-matter, Strings, etc. It has occurred to me that it is possible that a 2 dimensional saddle shaped Universe drawn in 4 dimensional space-time would, I believe, yield a Mr. Donut Universe. How's that for a real kicker! Graphed as a repeating equation in all four quadrants of a four dimensional graph, it could be expressed as a fractal or Julia set and would be in the shape of a beignet with a hole in the middle. Depending on how much space-time is curved it could even be A Mr. Donut Universe. How about that with your morning coffee! On to other related matters. These related matters continue with the evolving shape of the Universe and how it all ties together with the quantum construct.

Chapter 11

Genetics

For many years there have been two camps regarding life and it's either evolution or creation. This goes back to Charles Darwin and his book "Origin of Species". Unfortunately his tome should have been entitled "Variation of Species, through Natural Selection." Species Changes occur in a ring of interrelated forms that differ so much from either the starting point or the two ends differ so much they could be recognized for different species if the intervening or intermediary forms were not there or were extinct. The British herring gull and the black backed gull are an example. In Britain they often nest side by side but never interbreed. To look at them you would think they are two separate species, but they are merely two ends of a genetic ring of gulls that step by step change and can interbreed elsewhere with each other that extends around the world.

Paphiopedilums or Oriental lady slipper orchids could possibly be such a ring of forms in the plant world. Are there really 120 separate species of Paphiopedilums or are there 120 variations of a single species, or 20 varieties of six different ring species or one large interrelated ring species? Only D.N.A. analysis and time would answer that Question. "Mitochondrial Eve" was also an important clue that led me to this conclusion. SPECIES CHANGES OCCUR IN INDIVIDUALS, STEP BY STEP to form a ring of forms that the ends can no longer, or will no longer interbreed naturally. When the original form and intermediary forms die out, the members of the remaining forms become different species. I know this goes against the idea that a species is a clade of individuals that are genetically nearly identical, but hear me out if you will. How can this be you say? Let me explain that the branching changes of variation within a species changes so much that when the variations at each end of the species range are no longer able to interbreed or will no longer interbreed but may be able to breed with the original species. Then if the original species or intermediary varieties become extinct, the organisms at both ends of the spectrum of variation have become separate species. If the change is large enough over time, they become separate genera. If the changes are even larger over longer periods of time, they become separate phyla. This is probably what happened during the last billion years on Earth. So therefore, life is a continuing line of forms, changing gradually from one form into the next and if it wasn't for extinction there would and could be no species, as they are presently defined. Let me explain the exact method of speciation. For this we need to explore the world of amino acids, and genes. It has occurred to me that there are a number of methods of genetic change within a species to bring about enough changes to cause speciation and these methods are:

1. Changes in the genetic code, either during meiosis or mitosis (by changing a single amino acid; Adenine, Thymine, Guanine, or Cytosine, in a gene strand can make a gene accessible that was dormant or in-accessible.) This happens through inaccurate duplication of genetic material. A gene is not a straight strand of DNA; It has more twists, and turns than the greatest contortionist that ever lived. If a single DNA strand were pulled from a cell and straightened out it would stretch to a length of nearly six feet. That means that inside a cell much of the genetic information in a cell is buried within its own structure and inaccessible to the enzymes, RNA etc. of the cell. In order for a brain cell to be able to read the information for its function from the gene, the instructions MUST be available for the cell to read on the outside of the gene. Its a matter of location, location, accessibility, shape, size, entirety of the gene, and proper sequence. The sequencing is needed so that a liver cell only acts like a liver cell, not a brain cell, and a brain cell only acts like a brain cell not a liver cell. This supposition has just been verified by the Dec.2006 Issue of National Geographic on page 33 where it states:

> "Our DNA—specifically the 25,000 genes identified by the Human Genome Project—is now widely regarded as the instruction book for the human body. But genes themselves need instructions for what to do & when & where to do it. These instructions are not found in the letters of the DNA, BUT ON IT, an array of chemical markers, (isoleucine, leucine,) and switches, (Valine, Methionine, TAA, TAG, TGA, the last 3 are stops, the first two are starts) {My surmises are in parentheses} known collectively as the epigenome, that lie

along the length of the double helix of DNA. These epigenetic switches and markers in turn help turn on or off the expression of particular genes. Think of the epigenome as a complex (64 bit) software code, capable of inducing the DNA hardware into manufacturing an impressive array of proteins, cell types & individuals."

2. Looping, a form of genetic duplication, where the gene or part of the gene is produced twice instead of just once. This process is thought to have occurred in ungulates about 35 million years ago and mammals in general about 60 million years ago.
3. Genetic cross over, where part of one gene comes in contact with and becomes part of another gene and is so reproduced,
4. Polypoloidy, where all the genes of an organism are doubled through radiation or chemical means (such as colchicine, an alkaloid found in Autumn crocus.) not to be confused with genetic doubling, which is a different mechanism and produces a different effect. There are two forms of polypoloidy: autopopoloidy where a genome produces a double chromosome count for any one of a number of reasons including being acted upon by alkaloids such as cholhicine and allopolyoidy where two closely related genomes double their chromosome count, this is closely associated with genetic crossover.
5. Inversion is another form of genetic change. Inversion is where a segment or section of genetic code is inverted within the gene itself.
6. Translation is a form of genetic change where a section or segment of genetic code is transferred to another part of the same gene or a different part of another gene.
7. Folding, this changes the genetic accessibility and sequencing, thereby changing the genetic code. This process of genetic alteration goes on all the time, however for speciation to occur it MUST occur at specific locations, in a specific location, length, accessible, shape, size, and sequence. I call it Genetic Location, Access, Shape, Size, Entirety of the gene, & Sequence, G.L.A.S.S.E.S. for short.
8. Chromosome expansion and contraction are very similar to looping in that a part of a gene segment is repeated and so reproduced, genetic reduction is where a loop is formed but the loop is pinched off and the gene is thereby shortened.
9. Genetic Doubling, Which seems to have happened at least twice and maybe three times in the history of life on Earth, the first was the

doubling of the gene so that it could be read from either end and read the same. This probably happened some 800-900 million years ago.

The second time I believe it occurred was about 200 million years ago during age of dinosaurs, but it happened to mammals, not dinosaurs. The last time it occurred I believe was about 60-65 million years ago at the beginning of the age of Mammals. All of these forms of genetic change are occurring or have occurred in most genomes, some of them have occurred many times in the past. If the specific location for speciation is not changed, no change in species occurs. If the location that changes is specific for number of digits on their left hand and it changes the number from five to six, then the individual has six digits on the left hand. This happens with poly-dactyl cats. Most of the genetic changes are neutral and effect the individual neither one way or another. Some changes are Deadly (cystic fibrosis), some are specific to environmental conditions (Sickle cell anemia is an adaptation by Africans to the malaria parasite.)

IT IS NOT A RANDOM PROCESS !!

This process occurs because two amino acids, Valine; G.T.A. & G.T.C. and Methionine; A.T.G., indicate a start for the sequence following it to be read. With four amino acids and three positions, what we have is a 64 bit code written in base 4. This was a computer program first written some 900 million years ago. I believe Methionine is important in that it is one of two amino acids that contain Sulfur. The other Sulfur containing amino acid, Cysteine is a non-methylated form of the same amino acid. These amino acids can become methylated in the presence of such items as biotin and other methyl donors. As A.U.G., Methionine carries the start message for a ribosome to begin protein translation. Uracil is the Amino acid that takes the place of Thymine in Ribonucleic acid or R.N.A. This is significant, as during the creation of amino acid chains, Methionine is ALWAYS created first.

This also implies that it is also a start when instructions are being read by the cell. I believe both Valine and Methionine MUST be in sequence for the sequence to be accessible on the outside of the gene. I believe this pair of amino acids twists the strands in such away that they are forced to the outside of the genetic strand and are therefore readable. These methods of Speciation guarantee that natural selection will keep a species a species, and not create a new species except in individuals as I explained earlier. (Unfortunately the English language doesn't have a word for the concept I wish to express.) Certain individuals within a ring species may change enough to become the basis for further rings and eventually new species. Natural selection only makes individuals within a group fit into an environment better and provides for variation within a species. These methods have a six fold chemical safety factor built in to keep species within certain boundary limits as I showed earlier with G.L.A.S.S.E.S. The twenty amino acids that compromise the amino acids found in life are all left handed except guanine which is non-handed. Chemists have wondered why all of these amino acids are left handed, all sugars are right handed and most starches are right handed. There are a few exceptions, for instance inulin, the starch found in Jerusalem Artichokes is left handed. Since there are no right handed enzymes in most mammals, these starches are not digestible by most animals. Enzymes are almost always left handed. The reason these amino acids are left handed is because they have to be. Adenine Triphosphate, A.T.P., only occurs in the left handed version. It can not work in a right handed version. Enzymes act as keys to reduce complex sugars, complex starches, simple sugars and starches, proteins, amino acids and fats to simpler compounds and help the body to metabolize them.

So for an enzyme to dismantle a sugar or starch (which in effect is what it does) the enzyme must fit into the sugar or starch molecule and break it apart, which is why enzymes are left handed and sugars and starches are right handed. We will next explore the world of amino acids themselves. There most definitely was a "chemical evolution" that took place in the amino acids as Dana Kenyon first theorized in his book "Biological Predestination." I don't know how he missed it, but if you arranged the amino acids by chemical composition, with the smallest molecules coming first, Glycine and the largest molecules, Adenine and Guanine coming last, you will see a pattern emerging that splits into two groups, the purines, Adenine and Guanine and the pyrimidines, Cytosine, Thymine and Uracil. The amino acids change from one amino acid to another one mainly by four processes, methylation, the adding of a methyl group, CH_3, & CH_2 and ammoniation, the adding of an ammonia group, NH_2, & NH, acetylation, the adding of an acetyl group, CH_3OH, and the adding of a hydroxl group, OH. All of this is started with the replacing of a single Hydrogen atom with one of the above four groups. Sugars are formed in a similar fashion except they don't use the ammonia groups. Complex sugars are made by the process of removing a hydrogen atom from one or more simple sugars and a hydroxyl group from another sugar.

Starches are formed by removing a hydroxyl group from one complex sugar molecule and a hydrogen atom from another sugar molecule, making a chain of connected dehydrated sugars. Complex Starches are made by removing water from the simple starches into chains of complex starches. These starches can be as short as 4 starches to as long as several hundred. Cellulose takes the process a step farther by removing water from chains of starches in a similar fashion. Cellulose can contain Thousands of such dehydrated sugars, complex sugars, simple starches, & complex starches. Again the structure of these sugars, and starches, both simple and complex, and the structures of the cellulose are encoded on the OUTSIDE of the D.N.A. molecule. The processes of change or evolution of the amino acids as far as I can determine is as follows: I can not stress this enough, this is NOT strictly a Chemical evolution, Energy in the form of sunlight: infra-red, ultra-violet and x-ray radiation entered into and focused the direction of the reactions involved. This is where I believe the angle of deflection of the Stern-Gerlach Experiment is important, for if I am correct it would explain why the process works in only certain steps and in one direction only.

THEY ARE NOT RANDOM!!!

Glycine>	> Alanine	By the addition of a methyl group (CH_2) for a hydrogen atom at C^2
Alanine	> Cysteine	By the addition of a sulfur atom (S) and a hydrogen atom at C^3 for a hydrogen atom
Cysteine	> Methionine	By the addition of a methyl group (CH_2) for a hydrogen atom at C^3
Alanine	> Serine	By the addition of a Hydroxyl group (OH) for a hydrogen at C^3
Serine	> Threonine	By the addition of a acetyl group (CH_3OH) for a hydrogen at C^3
Threonine	> Aspartic Acid	By the addition of a hydroxyl and hydrogen atoms at C^4
Aspartic Acid	>Glutamic Acid	By the addition of a methane group (CH_3) at

The energy to form these chemical reactions was probably not supplied by the chemical environment, but by Quanta of energy (photons) supplied in the most part by the sun. Some energy may have been supplied by radioactivity, but I would say the largest amount of energy was supplied by the sun in the form of infrared, visible light, ultra violet and x-rays. The Nature of the Quantum construct allows for energy levels of Quanta to be absorbed in the construct at differing levels and therefore supplies differing levels of energy to initiate chemical reactions. Most of these reactions are in the photo-electric range and the Quanta of sunlight (infrared, Visible light, Ultra violet,) acts as both initiator and pattern allowing the reactions to proceed only in certain steps and in certain directions only.(Explanation bot. pg.) Strictly speaking the Quanta are not catalysts as they are consumed in the process, but they do act as catalysts in that they do speed up the reactions, they are specific in allowing only certain reactions to take place only in a certain order and a certain direction, they also convey information not only about the reactions themselves but also about the sunlight that caused the reactions.

The last two steps and the next two steps may be in a different order but I believe this section is where the amino acids separate into two different groups that eventually become the purines and the pyrimidines. From Aspartic Acid and asparginine to leucine and isoleucine the exact path to the purines and pyrimidines is not really clear, they do differentiate after this point and are unmistakable to produce purines and pyrimidines as end products. It also looks like these products go thru a process of de-hydroxylation (OH), dehydration, de-acetylation, de-methylation to become purines and pyrimidines.

Aspartic Acid	> Asparganine	addition of an amino group (NH_2) for a hydroxyl (OH) at C^4
Glutamic Acid	> Glutamine	addition of an amino group (NH_2) for a hydroxyl (OH) at C^5
Glutamine	> Valine	subtraction of amino group (NH_2) at C^5
Valine	> Proline	re-arrangement of structure, loss of two hydrogen, closing the ring, the first step in forming the pyrimidines.
Valine	> Leucine	addition of a methane group (CH_2) at C^4
Leucine	> Lysine	addition of an amino group (NH) at C^6
Leucine	> Isoleucine	different isomer of the same amino acid
Isoleucine	> Uracil	subtraction of a hydroxyl at C^1, the addition of an amino group (NH) at C^5, closing the ring

Uracil	> Thymine	addition of a methyl group at C^3, substitution of oxygen for a hydrogen at C^6 position (N)
Thymine	>Cytosine	changing placement of methyl group from C^3 to C^2
Lysine	> Arginine	addition of 2 amino groups (NH) at C^6
Proline	> Histidine	addition of a amino and a ethyl group
Alanine >	>Phenylalanine	added to a phenyl ring
Phenylanaline	>Tyrosine	insertion of Oxygen at C^4 in phenyl ring
Tyrosine	>Guanine	subtraction of hydroxyl from chain closing ring to form the purines
Guanine	> Adenine	adding an amino group (NH_2) for an oxygen
Guanine	> Tryptophan	re-arrangement of the external amino group (NH_2), addition of acetyl group (CH_3OH)

The sources for these changes can be traced back to the Quantum construct of the atom itself and the energy released by the sun in its quanta (photon) packets of infra-red, visible light, ultra violet and x-ray radiation. These two forces initiated the formation of amino acids, protein synthesis, RNA, and DNA. It's basically a form of organized energy transfer. It was NOT just a series of chemical reactions. I can not emphasize this enough. This I believe is where the exact angle of incidence and deflection is necessary and makes its presence known. If the angle is 23.5 °, as I suspect, then every thing else falls into place. Why 23.5 °? If you remember 23.5 ° is the exact axis of the earth's rotation. Why is that important? It would mean that, coupled with the magnetic field of the earth, there were only a few days each year that optimum conditions were right for each step of the chemical reaction to take place and since it can only progress in one direction because of the energy provided by the quanta of radiation, there was no break down of the material. There was a slow accumulation of the stuff of life. If I am correct, then once the collision that formed the moon and that knocked the Earth on a 23.5 ° axis, life on Earth was inevitable. I also believe the polar orbit of the quantum structure is tilted 23.5 °, that would be further proof of my ideas.

IT WAS NOT RANDOM!

Chapter 12

Form follows function $f(x) = x-u$; $\pm f(x) = x \pm u$; $f(x) = x^2 \pm u$

It was, in effect programmed by the radiant energy it received from the sun to progress in certain directions only. Life itself therefore was inevitable on the Earth. If you mapped out the above amino acids in three dimensions I'm sure you would get a derivation of a Julia set. A three dimensional construct of a two dimensional equation or formula. This should not surprise anyone, as we deal with three dimensional constructs in a number of fields, derived from two dimensional equations or formulae. In Mathematics, Chemistry, & Architecture people deal with three dimensional constructs in terms of two dimensional equations all the time. How is this any different, except in terms of scale? Are you beginning to see a pattern emerging? From The very, very, very small sizes to the scale of the Multi-verse, Every where you look one pattern keeps emerging.

JULIA SETS!

It's no coincidence! The Universe and Everything in it is part and parcel to the force that set it in motion.

A SINGLE MATHEMATICAL EQUATION THAT'S PART AND PARCEL TO EVERYTHING IN THE UNIVERSE.

It's built into everything in the Universe from amino acids and proteins to galaxy's and even the distribution of galaxy's in the Universe. It's even built into the atoms themselves. The Quantum construct itself is a Julia set, set in the micro-cosm of the sub-atomic and an atto-second time frame.

By James Walker

Made in United States
Orlando, FL
07 November 2024